Exploring the Fabric of Reality: Unveiling the Foundations of Existence

By James Rondepierre

Legal Page

Published by Meta2Physical Publishing

This book is a work of the author's imagination. Names, characters, places, and incidents are either the product of the author's imagination or used fictitiously. Any resemblance to actual events, locales, or persons, living or dead, is purely coincidental.

Editing and Design

This book was edited and designed with the support of KDP Publishing.

Meta2Physical Publishing

All artworks featured in this book are original creations by James Rondepierre and are available through the Meta2Physical Shop. The designs and imagery reflect the author's creative vision and serve to inspire and complement the themes of this work.

First Edition: 2025

Meta2Physical Shop products inspired by this book can be found at www.meta2physical.com.
For more information about the author, additional works, or to explore the Meta2Physical Shop, visit the website above.

Disclaimer

The ideas, perspectives, and information presented in this book are for informational and entertainment purposes only. The author and publisher make no representations or warranties with respect to the accuracy or completeness of the contents of this book and disclaim any liability for damages or losses that may result from its use. Readers are encouraged to exercise their own judgment and discretion.

Table of Contents

Chapter 1: Prologue to Existence

Delving into the Philosophical and Scientific
Inquiries about Existence

The Nature of Reality and the Human Quest for Meaning

In the hushed corridors of human thought, where the flickering candlelight of curiosity casts shadows upon the tapestry of time, profound questions about existence echo through the ages. These questions spark the flames of contemplation, inviting us to embark on a journey of discovery that transcends the boundaries of mere cognition. The human spirit, driven by an innate longing for understanding, delves into the deepest recesses of consciousness, reaching out to touch the very essence of reality itself.

From the earliest philosophers who pondered the origins of the cosmos, to the modern scientists who unravel the secrets of particles and galaxies, the pursuit of understanding existence has been a relentless endeavor. It's a quest that extends beyond the realm of empirical observation, delving into the realms of metaphysics, ethics, and even spirituality.

In the boundless expanse of the cosmos, where galaxies dance to the rhythm of gravity's embrace, we find ourselves suspended, both insignificantly small and profoundly significant. Are we mere witnesses to the unfolding universe, or does the intricate web of consciousness weave us into the fabric of creation? The interplay of the objective and the subjective, the tangible and the intangible, forms the crux of our inquiry.

As we contemplate the paradoxes of reality, we're faced with the duality of existence – the known and the unknown, the observed and the inferred. This chapter guides us through the labyrinth of ancient myths, philosophical treatises, and scientific revelations, shedding light on the fundamental nature of reality. We explore the nature of consciousness and its potential role as both observer and co-creator of the universe.

Moreover, we confront the intricate relationship between our quest for meaning and the very fabric of existence. Is meaning something inherent to the universe, waiting to be deciphered, or is it a construct we project onto the blank canvas of reality? The human journey, marked by its triumphs and struggles, its beauty and suffering, fuels the pursuit of significance in a seemingly indifferent cosmos.

As we venture forth, we'll encounter the brilliance of minds who've grappled with these questions across cultures and epochs. From Plato's allegorical cave to modern theories of quantum entanglement, from the Eastern philosophies of interconnectedness to the Western traditions of reason, the myriad perspectives converge to paint a multidimensional portrait of our existence.

Join us on this exploration of thought and wonderment, where the borders of disciplines blur and the boundaries of human understanding expand. Let us embrace the mysteries that beckon us further into the heart of existence, where the interplay of philosophy and science, of reason and awe, opens new doorways to the tapestry of reality.

In the hallowed halls of inquiry, humanity stands at the precipice of revelation, gazing into the abyss of existence and whispering, "Who are we, and what does it all mean?"

Chapter 2: The Cosmic Odyssey: Origins of the Universe

Big Bang Theory and the Birth of the Cosmos

Inflation, Cosmic Microwave Background Radiation, and the Formation of Galaxies

I n the vast theater of cosmic history, an overture of unimaginable energy heralded the birth of our universe. The Big Bang theory, a symphony of elemental forces and quantum fluctuations, set the stage for the grand narrative of existence itself. As we traverse the corridors of time, we encounter the concept of inflation, a breathtaking expansion that sculpted the cosmic landscape and paved the way for the formation of galaxies. Amidst this cosmic ballet, a whisper from the distant past—cosmic microwave background radiation—speaks to us, a cosmic echo of the universe's nascent moments.

In the crucible of primordial creation, a singularity—infinitesimally small and infinitely dense—held within itself the potential for all that would come to be. With a burst of unimaginable energy, space and time themselves unfurled, birthing matter and light in a cataclysmic event we now call the Big Bang. This pivotal moment, an overture to the cosmic drama, set the stage for the unfolding narrative of galaxies, stars, planets, and life.

Inflation: An Epoch of Cosmic Expansion

As the cosmic overture reverberated through the expanse, the universe underwent an astonishing transformation known as inflation. In the earliest moments after the Big Bang, the fabric of space itself expanded at an exponential rate, stretching infinitesimally small regions into the vast cosmos we behold today. Inflation sculpted the cosmic canvas, smoothing irregularities and providing the seedbed for the formation of galaxies and galaxy clusters. This epoch of rapid expansion, though fleeting in cosmic terms, left an indelible mark on the cosmos we inhabit.

Cosmic Microwave Background Radiation: Echoes of Creation

Billions of years later, as our understanding of the universe deepened, we stumbled upon a reverberating whisper from the distant past—the cosmic microwave background radiation. This faint radiation, bathing the universe in a gentle glow, serves as a cosmic time capsule, preserving the conditions of the universe when it was a mere infant. It offers us a glimpse into the primordial temperatures and densities that birthed galaxies and galactic clusters, validating the predictions of the Big Bang theory and confirming the remarkable journey from singularity to stars.

As we traverse the cosmic odyssey from the primal spark of the Big Bang to the gentle hum of cosmic background radiation, we witness the intricate dance of matter, energy, and space-time. The birth of galaxies, the formation of stars, and the emergence of the cosmos itself tell a tale of ceaseless transformation, a narrative etched upon the very heavens. This chapter guides us through the corridors of time, unveiling the grandeur of cosmic evolution and inviting us to contemplate our humble origins within the boundless expanse of the universe. In the vast cosmic symphony, the resonance of the Big Bang echoes through eternity, carrying with it the essence of all that is and all that shall be.

Chapter 3: The Dance of Particles: Subatomic World Revealed

Quantum Mechanics: The Peculiar Behavior of Particles at the Smallest Scales

Wave-Particle Duality, Uncertainty Principle, and the Role of Observation

In the realm of the infinitesimally small, the universe takes on a surrealist quality, where particles dance in a manner that defies our classical understanding of reality. This is the realm of quantum mechanics, a realm where the behavior of subatomic particles defies intuition and reshapes our perception of the universe. In this chapter, we delve into the enigmatic world of quantum physics, exploring the dual nature of particles, the profound implications of the uncertainty principle, and the curious role that observation plays in shaping reality at its most fundamental level.

Quantum Mechanics: Unveiling the Subatomic Realm

As we journey into the depths of matter, we encounter a reality that challenges our conventional notions. Quantum mechanics, a framework developed in the early 20th century, unravels the mysteries of particles on the tiniest scales. Particles exhibit behaviors that transcend classical physics – behaviors that are described by probability waves and complex equations rather than predictable trajectories. This chapter introduces us to the foundation of quantum mechanics, where particles become waves and the ordinary notions of causality are suspended.

Wave-Particle Duality: The Paradox of Presence

One of the most perplexing aspects of the quantum world is wave-particle duality. Subatomic entities like electrons and photons exhibit both particle-like and wave-like behaviors, depending on how they are observed or measured. This phenomenon challenges our understanding of objects as discrete entities and reveals the inherent ambiguity that characterizes the quantum realm. We navigate the intricacies of this duality, where particles oscillate between existence as localized entities and diffuse waves, defying our conventional understanding of reality.

Uncertainty Principle: The Limits of Knowledge

As we peer further into the subatomic realm, we encounter Werner Heisenberg's uncertainty principle—a cornerstone of quantum mechanics. This principle asserts that certain pairs of properties, like position and momentum, cannot be precisely known simultaneously. The more accurately we measure one property, the less accurately we can know the other. This inherent uncertainty sets a fundamental limit on our ability to know the properties of particles, revealing a profound epistemological boundary within the heart of physics.

The Role of Observation: Shaping Reality

Perhaps one of the most baffling aspects of quantum mechanics is the role that observation plays in determining the behavior of particles. The act of measurement collapses the wavefunction of a particle, transforming its probabilistic existence into a definite state. This concept has led to philosophical debates about the nature of reality and the interplay between the observer and the observed. This chapter delves into the philosophical implications of observation in quantum mechanics, raising questions about the nature of consciousness and the boundaries of scientific knowledge.

In the intricate dance of particles, quantum mechanics presents us with a world that challenges our intuitions and stretches the limits of our imagination. As we peer into the subatomic realm, we find a tapestry woven from uncertainty and paradox, inviting us to question the very nature of reality itself. This chapter guides us through the surreal landscapes of the quantum world, where particles and waves merge, and the observer becomes an integral part of the observed.

In the subatomic ballet, particles pirouette between existence and uncertainty, beckoning us to explore the quantum tapestry that shapes the essence of reality.

Chapter 4: The Fundamental Forces: Weaving the Cosmic Tapestry

Electromagnetism, Gravity, Strong and Weak Nuclear Forces

Grand Unified Theory and the Search for a Unified Force

In the intricate choreography of the cosmos, the fundamental forces act as cosmic choreographers, shaping the interactions that govern the behavior of matter and energy. These forces, distinct yet interwoven, underpin the very fabric of reality. From the ethereal pull of gravity to the vibrant dance of electromagnetism and the intimate realm of nuclear forces, this chapter unveils the intricacies of these forces and delves into the quest for a grand unified theory—a framework that seeks to weave these forces into a harmonious cosmic tapestry.

Electromagnetism: The Dance of Charges

Electromagnetism, the force that governs the behavior of charged particles, is a symphony of attraction and repulsion. As electrons and protons move through the universe, they engage in an intricate dance, creating electric and magnetic fields that shape the world around us. From the glow of stars to the pulsing current in our devices, electromagnetism weaves through every facet of existence, connecting particles across vast distances.

Gravity: The Cosmic Embrace

In the grand theater of space and time, gravity takes center stage as the force that shapes the paths of celestial bodies. Governed by Einstein's theory of general relativity, gravity bends spacetime itself, creating the warp and weft of the cosmic fabric. From the graceful orbits of planets to the majestic swirl of galaxies, gravity's gentle tug weaves the cosmos into a tapestry of cosmic harmony.

Strong and Weak Nuclear Forces: Binding the Subatomic World

At the heart of atomic nuclei, the strong and weak nuclear forces play a delicate duet, binding quarks and leptons into the symphonic structure of matter. The strong force, mediated by gluons, binds quarks within protons and neutrons, while the weak force governs processes like beta decay, shaping the landscape of nuclear transformations. These forces, acting on the smallest scales, carry profound implications for the evolution of stars, the creation of elements, and the delicate balance of cosmic life.

Grand Unified Theory: Seeking Cosmic Unity

The quest to unravel the tapestry of forces has led physicists on a journey toward unification—a quest for a grand unified theory. This theory seeks to merge electromagnetism, the weak nuclear force, and the strong nuclear force into a single elegant framework. Such a theory, if realized, could provide a deeper understanding of the fundamental forces and the conditions of the early universe. The elusive harmony of a grand unified theory beckons scientists to decipher the symphony of existence.

As we explore the interplay of these fundamental forces, we uncover the delicate balance that underlies the cosmos. The dance of electromagnetism, gravity, and the nuclear forces shapes galaxies, stars, and the very building blocks of matter. And as we stand on the precipice of a grand unified theory, we peer into the future of physics—a future where the cosmic tapestry might be woven even tighter, revealing deeper connections that bind the universe together.

In the grand interplay of forces, the cosmos weaves a symphony of interactions, harmonizing particles and galaxies, drawing the threads of existence into an intricate cosmic dance.

Chapter 5: Genesis of Life: From Chemicals to Cells

Origin of Life Theories and the Primordial Soup

Evolution, DNA, and the Emergence of Complex Organisms

In the vast theater of the cosmos, the emergence of life stands as one of the most remarkable acts. From the intricate dance of molecules to the symphony of cells, the story of life on Earth is a narrative of transformation and adaptation. This chapter delves into the captivating tale of life's genesis, exploring the diverse theories on how the spark of life ignited, and how it evolved over eons to give rise to the astounding diversity of organisms that grace our planet.

Origin of Life Theories: The Primordial Soup

In the quiet whispers of the early Earth, the origin of life remains an enigmatic question. The primordial soup theory, like an ancient recipe, suggests that the ingredients for life were present in the form of simple organic molecules. Pools of water, warmed by volcanic activity and struck by lightning, might have catalyzed the formation of amino acids—the building blocks of proteins. This chapter takes us back to the cradle of Earth, exploring the conditions that set the stage for the emergence of life from the raw elements of the planet.

Evolution and Natural Selection: The Dance of Adaptation

As life arose, it embarked on a grand journey of transformation—driven by the forces of evolution and natural selection. Through the eons, organisms competed, adapted, and diversified, forging the path from simple life forms to the dazzling array of species that populate the Earth. This chapter introduces us to the mechanism of natural selection, where beneficial traits are favored in the struggle for survival, and the tree of life branched into the myriad forms we observe today.

DNA: The Genetic Code of Life

At the heart of the intricate choreography of life lies DNA—a molecule that encodes the instructions for building and operating living organisms. The structure of DNA, a double helix, carries the genetic information that is passed from one generation to the next. As we decode the genetic alphabet, we uncover the profound story of inheritance and the mechanisms that drive the continuous evolution of life.

The Emergence of Complex Organisms: A Symphony of Cells

From single-celled organisms to the complex symphonies of multicellular life, the evolution of life has yielded a stunning array of forms and functions. Cells, the fundamental units of life, come together in intricate arrangements to form tissues, organs, and entire organisms. This chapter guides us through the evolution of complex life forms, celebrating the intricate dance of cells and the profound intricacies of biological systems.

As we journey through the epic of life's genesis, we find ourselves immersed in the symphony of transformation and adaptation. From the primordial stew of molecules to the dazzling diversity of life forms, the narrative of life on Earth is a testament to the creative power of evolution. This chapter invites us to marvel at the dance of life, from its humble origins to the flourishing ecosystems that grace our planet.

In the theater of Earth, the drama of life unfolds—a story of molecules and cells, adaptation and diversity, weaving the tapestry of existence with threads of evolution and vitality.

Chapter 6: The Mind Puzzle: Consciousness and Cognition

Neuroscience, Philosophy of Mind, and the Nature of Consciousness

The Hard Problem of Consciousness and the Quest to Understand Self-Awareness

In the intricate theater of the mind, the enigma of consciousness takes center stage—a riddle that has fascinated philosophers, scientists, and thinkers across epochs. What is the nature of our inner experience? How does the firing of neurons give rise to thoughts, emotions, and self-awareness? This chapter embarks on a journey into the depths of the mind, exploring the intertwining disciplines of neuroscience and philosophy to unravel the complexities of consciousness and the enduring quest to understand the mysteries of self-awareness.

Neuroscience and the Mapping of the Mind

Advancements in neuroscience have illuminated the intricate machinery of the brain, revealing the intricate networks of neurons that orchestrate our thoughts, emotions, and behaviors. The chapters of the brain's story—the cerebral cortex, the limbic system, the synapses—are unveiled, guiding us through the labyrinth of neural pathways that give rise to our conscious experiences. We delve into the neural dances that underlie perception, memory, and cognition, as we journey closer to the heart of consciousness itself.

Philosophy of Mind: The Theater of Thought

Concurrent with the scientific exploration of the brain is the rich tapestry of philosophical inquiry into the nature of the mind. The philosophy of mind grapples with questions that transcend mere physicality—questions about the nature of thought, the relationship between mind and body, and the mystery of subjective experience. Through the lenses of materialism, dualism, and emergentism, we navigate the philosophical landscapes that have shaped our understanding of the mind's inner workings.

The Hard Problem of Consciousness: A Philosophical Enigma

At the heart of the quest to understand consciousness lies the "hard problem." Proposed by philosopher David Chalmers, this problem delves into the realm of subjectivity—the experience of being. How can the firing of neurons, the dance of particles, give rise to the vivid and subjective experience of colors, sounds, and emotions? This chapter confronts the hard problem head-on, exploring theories of consciousness that seek to bridge the gap between neural activity and lived experience.

The Quest for Self-Awareness: The Inner Mirror

Among the riddles of consciousness, self-awareness stands as a jewel of the mind. What is the source of our sense of self? From mirror neurons to introspection, we unravel the layers of self-awareness—the ability to perceive ourselves as distinct entities with a continuous presence. Through psychology, neuroscience, and philosophy, we peer into the mirror of self-consciousness, seeking to comprehend the narrative of "I" that unfolds within our minds.

As we venture into the labyrinth of consciousness and cognition, we find ourselves amid the echoes of ancient inquiries and modern revelations. The interplay between neuroscience and philosophy reveals the intricate tapestry of the mind, where the symphony of neurons and the theater of thought converge. This chapter invites us to explore the profound conundrum of consciousness, a puzzle that remains both deeply personal and universally perplexing.

In the theater of the mind, the curtains rise on consciousness—a dazzling performance that interweaves neurons, thoughts, and self-awareness, inviting us to ponder the essence of being.

Chapter 7: Dimensions Unveiled: Space, Time, and Beyond

Einstein's Theory of Relativity and the Curvature of Spacetime

String Theory, Multiverses, and the Concept of Extra Dimensions

I n the grand tapestry of the cosmos, the interwoven fabric of space and time forms the stage upon which the drama of existence unfolds. Einstein's theory of relativity transformed our understanding of these dimensions, revealing the intimate connection between gravity and the curvature of spacetime. But could there be more dimensions lurking beyond our perception? This chapter embarks on a journey through the dimensions, from the curved paths of planets to the speculative realms of string theory and the mind-bending landscapes of multiverses.

Einstein's Theory of Relativity: Warping the Fabric of Reality

Einstein's theory of relativity, a monumental achievement of the 20th century, redefined our understanding of space and time. Special relativity revealed that space and time are intertwined, forming the spacetime continuum—a flexible fabric that warps in the presence of mass and energy. General relativity, on the other hand, unveiled the cosmic dance of gravity, where massive objects curve the paths of light and matter, shaping the orbits of planets and galaxies. This chapter guides us through the elegant equations that illuminate the curvature of spacetime and the interplay of matter and geometry.

String Theory: A Symphony of Vibrating Strings

As our quest for deeper understanding continues, we encounter string theory—an ambitious framework that seeks to unify the fundamental forces of nature and the laws of quantum mechanics into a single, harmonious structure. At the heart of string theory are tiny, vibrating strings—microscopic entities that dance through dimensions too small for our senses to perceive. We traverse the world of compactified dimensions and explore the diverse string theories that promise to unveil the ultimate symphony of the cosmos.

Multiverses: Beyond Our Cosmic Horizon

In the realm of speculative cosmology, the concept of multiverses emerges—a notion that suggests our universe might be just one of many, an infinitesimal bubble in a cosmic sea of parallel realities. From the inflationary multiverse to the brane cosmology of higher-dimensional spaces, we journey through theories that stretch our understanding of existence. This chapter navigates the landscape of multiverses, where the boundaries of our universe dissolve, giving rise to a tapestry of countless universes with unique properties and rules.

The Concept of Extra Dimensions: Hidden Realms

Could there be more to the dimensions than meets the eye? The idea of extra dimensions suggests that our familiar three dimensions of space might be accompanied by hidden dimensions that we cannot perceive. These dimensions, if they exist, could be tightly curled or compactified, offering new avenues of exploration for the fundamental forces of nature. This chapter leads us through the theoretical realms of extra dimensions, where the architecture of spacetime becomes an intricate and enigmatic symphony.

As we journey through the dimensions—both visible and speculative—we find ourselves at the threshold of reality's uncharted territories. From the curvature of spacetime to the vibrational harmonies of string theory, from the enigma of extra dimensions to the speculative landscapes of multiverses, this chapter invites us to explore the vast expanse of the cosmos and the mysteries that lie beyond the limits of our perception.

In the symphony of dimensions, space and time intertwine, bending and resonating in harmony with the hidden strings that weave the cosmic fabric of existence.

Chapter 8: Cosmic Evolution: Stars, Galaxies, and Beyond

Stellar Evolution, Supernovae, and Black Holes

Galaxy Formation, Supermassive Black Holes, and the Fate of the Universe

I n the vast cosmic theater, stars and galaxies form the constellations of existence—each one a luminary in the tapestry of cosmic evolution. This chapter unveils the remarkable story of how stars are born, live, and die, sculpting the galaxies that house them. As galaxies gather in cosmic clusters, we explore their formation and transformation. Amidst this cosmic dance, supermassive black holes emerge as enigmatic behemoths, shaping the destiny of galaxies. Our journey culminates with the grand canvas of the universe's fate.

Stellar Evolution: The Life Cycle of Stars

Stars, celestial luminaries born from the collapse of interstellar clouds, embark on a journey of transformation. From protostars to red giants, from supernovae to white dwarfs, stars progress through stages that span billions of years. This chapter follows the lives of stars, tracing their paths from fiery birth to celestial retirement. We explore the elements forged in the heart of stars, elements that form the building blocks of planets, life, and the cosmos itself.

Supernovae: Cosmic Crucibles

In the cosmic drama, supernovae are the spectacular crescendos that illuminate the heavens and scatter the seeds of creation. These explosive events mark the end of a star's life, releasing an astonishing amount of energy and forging heavy elements that enrich the cosmos. Supernovae bridge the gap between stellar lives and the universe's evolution, leaving behind remnants like neutron stars and black holes—a testament to the cosmic cycles of birth, life, and transformation.

Black Holes: The Cosmic Abyss

Among the most enigmatic entities in the cosmos are black holes—regions where gravity becomes an unstoppable force, warping spacetime to a point of no return. Stellar-mass black holes emerge from the remnants of massive stars, while supermassive black holes inhabit the hearts of galaxies. This chapter delves into the mysteries of black holes, exploring their formation, their influence on their surroundings, and their role in shaping the dynamics of cosmic evolution.

Galaxy Formation: Cosmic Collisions and Cosmic Assemblies

Galaxies, the building blocks of the universe, emerge from the primordial soup of matter and energy. Through gravitational interactions, galaxies gather in clusters and superclusters, forming cosmic cities of stars, gas, and dark matter. We journey through the cosmic ballet of galaxy formation, from the early universe to the intricate tapestry of cosmic structures we observe today. Amidst this dance, supermassive black holes take center stage, influencing the evolution of galaxies and leaving behind cosmic signatures.

The Fate of the Universe: Cosmic Destiny

As galaxies drift through the cosmic sea, we peer into the universe's fate—an expansive narrative that spans eons. Will the universe continue to expand forever, its galaxies drifting farther and farther apart? Or will gravity ultimately triumph, causing the cosmos to collapse in on itself? The cosmic destiny hinges on the balance between expansion and contraction, a story yet to be fully deciphered by scientists and thinkers alike.

As we navigate the tapestry of cosmic evolution, we encounter the stellar lives that illuminate the night sky and the galaxies that form constellations of cosmic beauty. The chapters of birth, transformation, and rebirth unfold on the canvas of spacetime, inviting us to ponder the grand narrative of the universe itself.

In the cosmic symphony, stars and galaxies compose the movements of cosmic evolution—a tale of birth, life, transformation, and the enigmatic abyss that shapes the destiny of the universe.

Chapter 9: The Human Experience: Emotions, Morality, and Society

Psychology, Emotions, and the Social Nature of Humanity

Ethical Frameworks, Cultural Differences, and the Search for a Universal Moral Code

In the vast landscape of existence, humanity emerges as a tapestry of emotions, thoughts, and relationships. This chapter delves into the intricate world of human psychology, exploring the emotional spectrum that colors our experiences. As social creatures, we form societies and cultures, weaving together ethics and morality. From the philosophical foundations of ethics to the nuances of cultural diversity, this chapter navigates the landscapes of the human experience.

Psychology: The Inner Landscape of Emotions

Within the realm of consciousness, emotions bloom like vibrant flowers, shaping our perceptions and interactions with the world. This section delves into the intricate fabric of human psychology, unraveling the emotional tapestry that spans from joy to sorrow, from fear to love. We explore the complexities of cognitive processes, the mysteries of the unconscious mind, and the ways in which emotions influence our behaviors and decisions.

The Social Nature of Humanity: Bonds and Communities

As social creatures, humans are intricately woven into the fabric of society. From families to communities, we navigate relationships that influence our identities, values, and actions. This section illuminates the social dynamics that shape our lives, examining the ways in which cultural norms, social hierarchies, and shared experiences influence our sense of self and our roles within the collective.

Ethical Frameworks: Pondering Right and Wrong

Ethics, the cornerstone of human morality, guides our actions and interactions in the world. From the utilitarian calculus of consequences to the deontological emphasis on duty, various ethical frameworks emerge, each offering insights into how we discern right from wrong. This chapter embarks on a philosophical journey through the moral landscape, exploring the evolution of ethical thought and the principles that underpin our decision-making.

Cultural Differences: The Tapestry of Perspectives

As humanity scatters across the globe, cultures emerge, each with its own set of values, beliefs, and norms. These cultural differences paint a vibrant tapestry of perspectives, inviting us to contemplate the diverse ways in which people perceive the world and navigate moral dilemmas. We traverse the globe, exploring the nuances of cultural relativism, the challenges of moral universalism, and the interplay between cultural traditions and evolving ethical standards.

The Search for a Universal Moral Code: Ethics Beyond Boundaries

In the midst of cultural diversity, a quest for a universal moral code emerges—a set of ethical principles that transcends cultural boundaries and reflects the shared essence of humanity. This chapter navigates the currents of moral philosophy, exploring the idea of human rights, the search for common ground in an interconnected world, and the potential for ethical systems that bridge cultural divides.

As we journey through the landscapes of the human experience, we encounter the intricate interplay of emotions, relationships, ethics, and cultures. From the inner workings of the mind to the complex tapestry of societies, this chapter invites us to explore the rich mosaic that defines the essence of being human.

In the mosaic of humanity, emotions color our experiences, ethical frameworks guide our actions, and cultural diversity shapes our perspectives, all woven together in the intricate tapestry of the human experience.

Chapter 10: Technological Epoch: From Fire to AI

Technological Advancements Throughout Human History

Artificial Intelligence, Transhumanism, and the Ethical Challenges of Advanced Technology

In the annals of human progress, the story of technology unfolds as a testament to our ingenuity and ambition. From the primitive mastery of fire to the intricate world of Artificial Intelligence (AI), this chapter chronicles the transformative journey of human innovation. As we peer into the realm of AI and the potential for transhumanism, we confront profound ethical challenges that beckon us to consider the impact of advanced technology on the fabric of society and the essence of humanity itself.

Technological Advancements: From Fire to the Digital Age

The human odyssey of technological advancement stretches across millennia, marked by pivotal milestones that shape the course of civilization. We traverse the landscape of innovation, from the mastery of agriculture to the revolution of industry, from the harnessing of electricity to the dawn of the digital age. Each step in the journey redefines the boundaries of what is possible, offering new tools and possibilities that reshape our interactions with the world.

Artificial Intelligence: The Frontier of Innovation

In the modern era, the horizon of technological innovation extends into the realm of Artificial Intelligence—an endeavor that seeks to imbue machines with cognitive capabilities akin to human intelligence. From machine learning to neural networks, AI systems have transformed industries, revolutionized data analysis, and opened new frontiers in scientific research. This section explores the evolution of AI, the challenges of creating intelligent machines, and the potential for AI to shape the future of humanity.

Transhumanism: The Evolution of Human Potential

As technology extends its reach into the human experience, the concept of transhumanism emerges—a movement that envisions a future where technology is integrated with biology, allowing humans to transcend their current limitations. From genetic engineering to brain-computer interfaces, transhumanism opens the door to a world of possibilities, from enhanced cognitive abilities to increased longevity. We contemplate the ethical implications of augmenting human capabilities and the balance between embracing progress and preserving human identity.

Ethical Challenges of Advanced Technology: A Moral Crossroads

As we tread the path of technological advancement, we encounter ethical crossroads that demand careful consideration. From the ethical dilemmas posed by autonomous AI to the potential consequences of altering the human genome, these challenges test our values, principles, and visions of the future. This section navigates the labyrinth of ethical concerns, addressing questions about privacy, control, equality, and the preservation of human dignity amidst rapid technological change.

As we reflect on the tapestry of technology, from the primal mastery of fire to the frontiers of AI and transhumanism, we find ourselves at a pivotal juncture in human history. The innovations that shape our world offer boundless possibilities, yet they also challenge us to grapple with the implications of our creations. This chapter invites us to ponder the role of technology in shaping our identity, our values, and the course of humanity's journey.

In the realm of technology, from fire's ancient glow to AI's futuristic potential, humanity stands at the crossroads of innovation and ethics—a threshold where progress and responsibility intersect.

Chapter 11: Harmony with Nature: Ecology and Sustainability

Ecosystems, Biodiversity, and the Interconnectedness of Life

Climate Change, Environmental Conservation, and the Human Impact on the Planet

I n the intricate web of existence, Earth's ecosystems form a delicate balance, sustaining the diverse tapestry of life. This chapter delves into the realm of ecology, exploring the intricate connections that weave life forms into intricate ecosystems. As humanity stands as both custodian and disruptor of these systems, we confront the challenges of climate change, environmental conservation, and the ethical responsibilities that come with our role in the natural world.

Ecosystems and Biodiversity: The Dance of Life

Within the embrace of Earth's landscapes, ecosystems flourish—a symphony of interdependent life forms interacting with their environment. From the lush rainforests to the arid deserts, ecosystems showcase the intricate dance of producers, consumers, and decomposers, each thread weaving a unique role into the ecological fabric. This section navigates the tapestry of biodiversity, where every species contributes to the stability and resilience of the planet's delicate systems.

Interconnectedness of Life: From

Microbes to Biomes

In the grand theater of ecology, the interconnectedness of life resonates through every level of organization. From microscopic microbes to expansive biomes, the harmonious web of interactions shapes the flows of energy and matter across the Earth. We explore the complex networks that sustain life, from nutrient cycles and food webs to the profound ways in which species influence one another's survival.

Climate Change: A Global Challenge

As humanity's presence on Earth grows, so does our impact on the planet's climate. The chapter delves into the sobering reality of climate change, exploring the scientific evidence of rising temperatures, sea level rise, and extreme weather events. We contemplate the intricate feedback loops that drive climate dynamics and the ethical responsibility to address this global challenge—by transitioning to sustainable practices, reducing greenhouse gas emissions, and preserving the health of our planet.

Environmental Conservation: Guardians of the Earth

In the face of environmental challenges, a movement for conservation and sustainability emerges—a call to protect the natural world and ensure its health for future generations. From the establishment of protected areas to the push for sustainable practices in agriculture and industry, conservationists work to safeguard biodiversity, restore ecosystems, and mitigate the impacts of human activity. This section celebrates the efforts of those who champion the cause of environmental stewardship.

As we explore the chapters of ecology and sustainability, we confront the profound implications of our actions on the planet. From the intricate interplay of life within ecosystems to the urgent need for climate action, this chapter invites us to embrace our roles as custodians of Earth and to contemplate the harmonious relationship we must forge with the natural world.

In the symphony of ecology, Earth's ecosystems play a harmonious melody—a testament to the interconnectedness of life and the ethical responsibility to safeguard the beauty and balance of our planet.

Chapter 12: Unifying Knowledge: Science, Spirituality, and Philosophy

Bridging the Gap Between Scientific Understanding and Spiritual Beliefs

Eastern and Western Philosophies, Mysticism, and the Quest for Unity

In the grand exploration of existence, the realms of science, spirituality, and philosophy unfold as diverse paths toward understanding the mysteries of the universe. This chapter delves into the endeavor to bridge the gap between empirical knowledge and spiritual insights, acknowledging the common quest for truth that unites these disciplines. As Eastern and Western philosophies intermingle, we explore the realms of mysticism, contemplation, and the timeless search for unity in the tapestry of existence.

Bridging Science and Spirituality: The Quest for Truth

In the dance of human inquiry, science and spirituality have often been perceived as separate realms—one focused on empirical evidence, the other on matters of faith and transcendence. Yet, throughout history, thinkers and seekers have explored the intersections of these realms, seeking to harmonize the empirical and the spiritual. This section navigates the terrain of quantum mysticism, the dialogue between science and religion, and the potential for a holistic worldview that bridges the apparent gaps.

Eastern and Western Philosophies: Diverse Paths to Understanding

Across cultures and eras, human thought has given rise to a multitude of philosophical perspectives—each offering unique insights into the nature of reality, existence, and the human experience. We journey through Eastern philosophies that emphasize interconnectedness, mindfulness, and the exploration of inner realms. We also explore Western philosophies, which probe the nature of knowledge, morality, and the intricacies of the human condition. Together, these perspectives weave a rich tapestry of wisdom that invites us to contemplate the profound questions of life.

Mysticism: Exploring the Transcendent

Mysticism, a thread that weaves through the fabric of spirituality, seeks to encounter the ineffable and transcendental aspects of reality. Across traditions, mystics have ventured into the realm of direct experience, seeking union with the divine or ultimate truth. This section delves into the mystical journeys of individuals who have plumbed the depths of consciousness, transcended the boundaries of the ego, and touched upon the unity that underlies all existence.

The Quest for Unity: Threads of Oneness

At the heart of both scientific inquiry and spiritual exploration lies the quest for unity—a search for the underlying principles that connect all aspects of existence. Whether through the unification of fundamental forces or the realization of spiritual enlightenment, humans have long sought to unravel the tapestry of unity that weaves through the cosmos. This chapter contemplates the bridges between scientific theories like the theory of everything and spiritual notions of interconnectedness, inviting us to ponder the ultimate nature of reality.

As we navigate the realms of science, spirituality, and philosophy, we discover that these disciplines converge in the shared human endeavor to understand our place in the cosmos. In the tapestry of existence, the search for truth is a common thread that binds thinkers, seekers, and explorers across cultures and epochs. This chapter invites us to contemplate the diverse paths that lead to understanding and the potential for a unified perspective that transcends boundaries.

In the symphony of human inquiry, science, spirituality, and philosophy harmonize—a chorus that seeks to grasp the mysteries of existence, weaving threads of insight and unity into the fabric of understanding.

Chapter 13: The Eternal Enigma: Death and Beyond

Cultural Perspectives on Death and the Afterlife

Immortality, Digital Consciousness, and the Ethical Implications of Extending Life

In the grand narrative of existence, the enigma of death stands as a universal and profound experience—a threshold that sparks contemplation, fear, and wonder across cultures and time. This chapter delves into the diverse perspectives on death and the afterlife, exploring the various ways in which humanity grapples with the mysteries that lie beyond. As technology advances, the concept of immortality takes new forms, prompting us to confront ethical questions about the extension of life and the potential for digital consciousness.

Cultural Perspectives on Death: A Tapestry of Beliefs

Across cultures and civilizations, death has been understood and interpreted in a multitude of ways. From the ancient practices of honoring ancestors to the modern rituals that bid farewell to loved ones, this section delves into the cultural dimensions of death. We explore the concepts of reincarnation, resurrection, and the eternal cycle of existence, tracing the threads of belief that weave through the fabric of human spirituality.

The Afterlife: Visions of Beyond

The concept of an afterlife has intrigued humanity for centuries, offering visions of a realm beyond mortality. From heaven and hell to the realms of the spirit world, different traditions offer diverse perspectives on what lies beyond the veil of death. This section contemplates the notions of souls, the journey of the departed, and the possibility of continued existence in realms that transcend the physical.

Immortality: Quests for Eternal Life

Advances in technology and science prompt us to consider the possibility of immortality—a life that extends beyond the natural boundaries of existence. From the alchemical quests for the elixir of life to modern scientific explorations of life extension, the concept of immortality takes on new forms. This section explores the ethical implications of immortality, raising questions about the quality of life, the balance between longevity and overpopulation, and the potential consequences of defying the natural order.

Digital Consciousness: The Boundaries of Being

In the era of rapid technological advancement, the concept of extending life takes a futuristic twist—toward the preservation of consciousness in digital form. Could the human mind be uploaded into computers, granting a form of digital immortality? This section contemplates the possibilities and challenges of transferring consciousness to the digital realm, addressing questions about identity, autonomy, and the ethical dilemmas that arise when humans blur the boundaries between biology and technology.

As we explore the realm beyond mortality, we encounter a complex interplay of belief, science, and ethics. From cultural perspectives on death to the enigmatic visions of the afterlife, from the quest for immortality to the potential for digital consciousness, this chapter invites us to contemplate the profound questions that lie at the heart of the eternal enigma.

In the enigma of death and beyond, humanity confronts the mysteries of existence—a journey that spans cultural beliefs, technological frontiers, and ethical considerations, as we seek to fathom the uncharted territories of the afterlife and immortality.

Chapter 14: The Quest for Truth: Ethics of Knowledge and Wisdom

The Pursuit of Truth and the Ethical Responsibilities of Knowledge

Wisdom, Humility, and the Balance Between Progress and Responsibility

In the labyrinth of human inquiry, the pursuit of truth stands as a noble endeavor—a journey that unravels the mysteries of existence and shapes the course of human progress. This chapter explores the ethical dimensions of knowledge, acknowledging the responsibilities that come with the acquisition and dissemination of truth. As we delve into the realms of wisdom, humility, and the delicate balance between advancing understanding and embracing ethical stewardship, we contemplate the tapestry that emerges when knowledge intertwines with ethics.

The Pursuit of Truth: Ethical Responsibility

As seekers of knowledge, we embark on a quest for truth that illuminates the pathways of understanding. But with knowledge comes responsibility—a recognition that the truths we uncover have the power to shape societies, individuals, and the course of history. This section navigates the ethical considerations that accompany the pursuit of knowledge, addressing questions about transparency, accountability, and the impact of information on humanity's collective narrative.

Wisdom: The Beacon of Guiding Insight

Beyond the acquisition of knowledge lies the virtue of wisdom—a quality that transcends mere information, embodying the ability to discern, judge, and act with ethical integrity. Wisdom guides us in understanding not just the "what" of knowledge, but also the "how" and "why." This section explores the cultivation of wisdom, drawing insights from philosophical traditions, historical narratives, and the personal journeys of those who seek to apply knowledge in ways that uplift and empower.

Humility in the Face of Uncertainty

Amidst the vast expanse of knowledge, there lies a humbling truth—the limits of human understanding. This section invites us to embrace humility as we navigate the frontiers of knowledge, acknowledging that our grasp of reality is ever-evolving and that there are dimensions that lie beyond our current reach. We explore the balance between the excitement of discovery and the humility required to acknowledge the mysteries that persist.

Progress and Responsibility: Navigating the Ethical Terrain

In the march of progress, the ethical terrain becomes increasingly complex. As technology, science, and knowledge advance, so do the ethical dilemmas that arise. This section navigates the landscapes of genetic engineering, artificial intelligence, and other frontiers of innovation, inviting us to contemplate the balance between pushing the boundaries of understanding and preserving the well-being of humanity and the planet.

As we embark on the quest for truth and wisdom, we find ourselves at a crossroads where knowledge and ethics intersect.

From the pursuit of truth to the cultivation of wisdom, from humility in the face of uncertainty to the ethical responsibilities of progress, this chapter invites us to reflect on the intricate interplay between knowledge and ethics, guiding our journey toward a future that is both enlightened and responsible.

In the pursuit of truth and wisdom, knowledge and ethics converge—a symbiosis that shapes our understanding and guides us toward a future where progress and responsibility walk hand in hand.

Afterword

Reflecting on Our Journey through Existence

Dearest Readers,

As we bring our exploration of existence to a close, we find ourselves at a juncture of reflection—a moment to contemplate the depths we have plumbed, the horizons we have glimpsed, and the tapestry of knowledge and wonder we have woven together. The journey through these pages has been a voyage through the vast expanse of human thought, an odyssey that has taken us from the origins of the cosmos to the depths of consciousness, from the intricacies of particles to the grand narrative of existence itself.

Throughout the chapters that have unfolded, we have delved into the realms of science, philosophy, spirituality, ethics, and more. We have pondered the mysteries of reality, explored the nuances of human experience, and sought to bridge the gaps between disciplines and perspectives. We have traced the threads of knowledge and wisdom, recognizing the common threads that weave through cultures, epochs, and the tapestry of humanity.

But as we close this book, we must acknowledge that our exploration is just a glimpse—a glimpse into a universe that is infinitely vast, a glimpse into questions that are boundless in their depth, and a glimpse into the unending journey of understanding that stretches before us. The questions we have asked and the answers we have found are but signposts along a path that leads to ever more questions, ever more insights, and ever more discoveries.

Our journey through existence is a collective endeavor, a symphony of minds, cultures, and perspectives. It is an ode to the human spirit—the spirit of inquiry, curiosity, and wonder that drives us to explore the mysteries that surround us. It is a reminder that as we seek to understand the world around us, we also seek to understand ourselves, our place in the cosmos, and the connections that bind us to each other and to all of existence.

As you close this book and continue on your own journey, remember that the quest for truth, knowledge, and wisdom is a lifelong pursuit. Embrace the mysteries that remain, for they are the sparks that ignite our curiosity and drive us forward. Continue to explore, question, and seek understanding, for it is in the pursuit of knowledge that we find both purpose and fulfillment.

May this exploration of existence serve as a guide, a companion, and an inspiration on your own journey through the uncharted territories of knowledge, discovery, and understanding. As you navigate the intricacies of reality, may you find wonder in the smallest details and awe in the grandest expanses. And may your journey be filled with the joy of discovery, the humility of learning, and the profound sense of connection that comes from contemplating the wondrous tapestry of existence.

With gratitude and anticipation for the chapters yet to be written,

James Rondepierre w/ ChatGPT
Author / Artist

Conclusion: Embracing the Mysteries that Remain

In our journey through the intricacies of existence and the fabric of reality, we have delved into the profound depths of scientific inquiry, philosophical contemplation, and metaphysical speculation. Along this odyssey, we have encountered fascinating theories and glimpsed the edges of comprehension, yet the enigma of existence persists. The more we unravel, the more mysteries seem to emerge, challenging our understanding and beckoning us to explore further.

As we conclude this exploration, we find ourselves standing at the threshold of the unknown, surrounded by the tantalizing allure of uncharted territories. The foundations of existence, though illuminated by the torch of human inquiry, cast shadows that invite us to embrace the uncertainties that remain. It is in this embrace of the mysteries that we discover the true essence of our quest—a perpetual journey toward understanding, fueled by curiosity and tempered by humility.

The journey through the fabric of reality is ongoing, and the tapestry of existence continues to unfurl before us. As we navigate the intricate threads of science, philosophy, and consciousness, we must acknowledge that some questions may forever elude our grasp. Yet, it is precisely in this acknowledgment that we find the spark that propels us forward, inspiring new questions, new explorations, and new revelations.

As we embark on the next chapter of our collective inquiry, let us carry with us the spirit of curiosity, the courage to confront the unknown, and the humility to recognize the limits of our understanding. The fabric of reality remains a rich tapestry, woven with threads of complexity and wonder. In embracing the mysteries that remain, we continue to unravel the secrets of existence and contribute to the ongoing dialogue that defines our intellectual and spiritual journey.

Certificate of Purchase

This is to certify that

_____

has acquired a copy of the work, *Exploring the Fabric of Reality: Unveiling the Foundations of Existence,* authored by James Rondepierre. This certificate acknowledges your participation in the timeless pursuit of knowledge, the exploration of profound questions, and the shared quest for deeper understanding of the nature of existence. By obtaining this work, you have contributed to the ongoing dialogue surrounding science, philosophy, and metaphysics, joining a global community of thinkers and seekers of truth.

Presented to:

_____

Presented by:

_____ (if given as a gift)

Date of Purchase:

_____

Personal Message (Optional):

_____

May this book inspire your journey of discovery, spark meaningful reflection, and deepen your connection with the mysteries of existence. Your support of this work helps keep the spirit of curiosity and learning alive for generations to come.

_____

With gratitude,

James Rondepierre

About the Author

James Rondepierre is a visionary author, spiritual teacher, artist, and thinker whose works have touched the hearts and minds of readers and seekers worldwide. Born in Springfield, Massachusetts, and raised in the suburbs of Philadelphia, James has spent a lifetime exploring the vast dimensions of human experience, spirituality, and creativity. At the time of publishing this book, he resides in North Carolina, where the serene beauty of nature continues to inspire his journey.

A Life of Exploration and Connection

James has traveled extensively, visiting 44 states across the United States, marveling at the natural wonders of national parks, and delving into the cultural richness of the places he's explored. His international travels have taken him through Mexico and beyond, broadening his understanding of humanity's interconnectedness. These experiences fuel his creative works, imbuing them with themes of exploration, unity, and transcendence.

Passion and Spirituality

James's life is defined by his deep love and passion for spiritual growth, creativity, and connection. His writings and teachings reflect his profound appreciation for human existence and his unwavering belief in the power of love, compassion, and unity. A deeply spiritual person, he incorporates the wisdom of diverse traditions into his works, offering readers universal truths that inspire self-discovery and enlightenment.

He is a devoted humanitarian, guided by his faith and love for life. With a heart that seeks to uplift and a soul attuned to the beauty of the world, James finds joy in the simple and profound—whether it's the serenity of a beach, the laughter of loved ones, or the vibrant creativity that flows through his art and writing.

Transformative Literary Works

James Rondepierre's books are celebrated for their depth, clarity, and transformative power. His works are available on Amazon in Kindle, paperback, and hardcover formats, with Audible editions soon joining the collection. Each book is crafted to guide readers on their journey toward understanding, personal growth, and spiritual awakening.

Artistic Mastery

As the creative force behind *Meta2Physical Shop*, James has pioneered a unique artistic style that blends sacred geometry, celestial symbolism, and infinitely tiling designs. His artwork is available on platforms like Redbubble, where his creations transform ordinary items into extraordinary expressions of beauty and meaning. His passion for creativity is evident in every piece, reflecting the themes of his literary works.

A Legacy of Light

Through his books and art, James Rondepierre invites readers to embark on transformative journeys of self-discovery, spirituality, and creativity. His dedication to exploring the infinite potential of human consciousness and the mysteries of existence ensures that his work will continue to inspire for generations to come.

Explore James's literary and artistic works at:

Meta2Physical.Redbubble.com

Thank you for being part of this journey, and may the light of these stories guide your path as they have guided his.

Epilogue: The Unending Journey

Embracing the Mysteries That Continue to Elude Human Comprehension

E ncouragement to Continue Exploring, Questioning, and Expanding Our Understanding of Existence

As we stand on the threshold of understanding, gazing into the vast expanse of existence, we are humbled by the mysteries that continue to elude our grasp. The journey through the preceding chapters has been a voyage of exploration, contemplation, and discovery—a journey that has illuminated the diverse facets of existence that shape our lives and our understanding of the cosmos.

In the symphony of existence, there are notes that remain beyond our hearing, harmonies that transcend our senses. The tapestry of reality is woven with threads of wonder that remind us of the vastness of the unknown. As much as we have uncovered and understood, there are horizons that beckon us to venture further, questions that invite us to seek answers beyond our current understanding.

The pursuit of knowledge, truth, and wisdom is a ceaseless endeavor, a journey that unfolds with each step we take. It is a testament to the insatiable curiosity that resides within us, urging us to explore the realms of science, philosophy, spirituality, and beyond. The mysteries that elude us are not barriers but invitations—to continue exploring, questioning, and expanding the boundaries of our understanding.

As we navigate the unending journey, we are called to embrace the humility that comes with knowing that there is always more to discover, more to learn, more to explore. Let us celebrate the questions as much as the answers, for each inquiry is a step closer to unraveling the intricate tapestry of existence. The pursuit of knowledge is a collective effort, a dance of minds and souls, a symphony of human endeavor that spans across time and space.

In the chapters that have unfolded, we have journeyed through the cosmos and into the depths of human experience. We have explored the realms of science, spirituality, ethics, and more. And as we close this chapter of the narrative, we leave you with an open invitation—to embrace the mysteries, to cherish the journey, and to embark on the unending quest to understand the essence of existence.

In the grand tapestry of the universe, as in the chapters of this exploration, the journey continues—an eternal voyage of discovery, contemplation, and the unceasing pursuit of understanding.

"Exploring the Fabric of Reality" aims to present a comprehensive and thought-provoking journey through the most important topics that shape our existence. By weaving together scientific knowledge, philosophical insights, and cultural perspectives, this book invites readers to contemplate the intricate tapestry of reality and their place within it.

List of Books by James Rondepierre

(Available on Amazon for Kindle, on Audible, in paperback, & hardcover formats, and some in different language translations!)

1. **The Nexus of Worlds: With Bonus Content**
   Embark on a mesmerizing journey through interconnected realms in *The Nexus of Worlds*. This gripping tale unravels the mysteries of parallel universes and invites readers to dive deeper with bonus content for an enriched experience.

2. **Mastering Luck: Comprehensive Guide to Lottery and Gaming Strategy**
   Discover strategies for navigating the intricate world of lottery and gaming with *Mastering Luck*. This comprehensive guide reveals secrets behind mastering the elusive force of luck.

3. **Exploring the Infinite Realm: Unveiling the Mysteries of Dreams**
   Exploring the Infinite Realm takes readers on an enchanting journey through the profound mysteries of dreams, delving into the limitless possibilities of the dreamworld.

4. **Exploring Karma: Understanding the Law of Cause and Effect**
   Gain insights into the workings of karma with *Exploring Karma*. This book offers a transformative journey into the universal law of cause and effect, guiding personal growth and understanding.

5. **Harvesting American Ginseng: A Comprehensive Guide**
   Delve into the world of American Ginseng with *Harvesting American Ginseng*. This guide provides practical insights into harvesting and explores the cultural and medicinal significance of this revered plant.

6. **The Precision Prognosticator: Navigating the Path to Accurate Future Prediction**
   Step into the realm of precision predictions with *The Precision Prognosticator*. This guide offers valuable insights into foreseeing the future with accuracy and understanding intuitive abilities.

7. **Embracing Serenity: Navigating Life's Challenges with Peace, Love, and Happiness**

In *Embracing Serenity*, readers are invited to navigate life's challenges with grace, peace, and love. This exploration serves as a guide to finding inner peace and happiness.

8. **The Subliminal Brilliance Blueprint: Unleashing Your Hidden Superpowers in Higher Dimensions**
Uncover the blueprint of subliminal brilliance with *The Subliminal Brilliance Blueprint*. This guide explores untapped potential within higher dimensions, offering a roadmap to unlocking hidden superpowers.

9. **Veil of the Night: Unveiling the Vampiric Nature of Humanity**
Veil of the Night invites readers to unravel the mysteries of the night and explore the vampiric nature of humanity. This tale blends the supernatural with the human experience.

10. **Transcending Realities: A Holistic Exploration of Consciousness, Shifting Realities, and Self-Realization: Part I**
Transcending Realities: Part I takes readers on a profound journey through consciousness, shifting realities, and self-realization, offering a multi-faceted perspective on existence.

11. **The Quantum Wealth Code: Unleashing Multiversal Prosperity**
Unlock the quantum wealth code with *The Quantum Wealth Code*. This guide provides insights into prospering across multiple universes and unlocking abundance in various aspects of life.

12. **Whispers of the Soul: Love, Sex, and the Sacred Union**
Delve into realms of love, sex, and spirituality with *Whispers of the Soul*. This exploration offers deep insights into the sacred union of souls, contemplating the deeper dimensions of human connection.

13. **The Symphony of Joy: Embracing Life's Grand Design: Includes Bonus Content!**
The Symphony of Joy invites readers to embrace life's grand design. This

edition includes bonus content, adding extra inspiration and joy to the exploration of existence's beauty.

14. **Rediscovering The World: A Journey through Anosmia**
    Embark on a sensory journey with *Rediscovering The World*. This exploration provides a unique perspective on the world through anosmia, offering a captivating and introspective experience.

15. **Evolving Unity: A Journey to Enrich All Existence, Elevate All Life, and Uplift Humanity**
    Evolving Unity beckons readers on a transformative journey to enrich existence, elevate life, and uplift humanity, serving as a guide for unity and collective growth.

16. **100 of the Greatest Stories Ever Told**
    Immerse yourself in *100 of the Greatest Stories Ever Told*. This collection promises a journey through captivating narratives spanning different genres and eras.

17. **Cosmic Wealth: Unleashing the Mystical Forces of Prosperity and Abundance**
    Unleash cosmic wealth with *Cosmic Wealth*. This guide provides a roadmap to attracting prosperity and abundance by tapping into mystical forces.

18. **The Modern Day Holy Bible**
    Explore spirituality in the modern era with *The Modern Day Holy Bible*. This perspective on timeless wisdom invites readers to contemplate profound teachings.

19. **The Radiance Within: Embracing the Joys, Pleasures, and Purpose of Human Existence**
    The Radiance Within invites readers to embrace the joys, pleasures, and

purpose of human existence, encouraging self-discovery and a deeper connection with life.

20.   **Ethereal Bonds: Love Unveiled**
Unveil the ethereal bonds of love with *Ethereal Bonds*. This exploration delves into the mysteries and beauty of love, reflecting on the transformative power of human connection.

21. **100 Stories**
Immerse yourself in *100 Stories*. This collection offers a tapestry of narratives spanning genres and themes for a rich and engaging reading experience.

22. **The Symphony of Infinite Wisdom**
Dive into the celestial chronicles with *The Symphony of Infinite Wisdom*. This book offers profound insights and timeless wisdom for a deeper understanding of life's mysteries.

23. **Miracles: Unraveling the Extraordinary Mystery of Divine Intervention**
Miracles unravels the mystery of divine intervention, inviting readers to contemplate the extraordinary occurrences that defy explanation and glimpse the miraculous in everyday life.

24.   **Healing Lupus Naturally: A Holistic Approach**
Discover holistic approaches to healing lupus with *Healing Lupus Naturally*. This guide provides a holistic perspective on health, offering hope and practical strategies for those with autoimmune conditions.

25. **Transcending Realities: A Holistic Exploration of Consciousness, Shifting Realities, and Self-Realization: Part II**
Transcending Realities: Part II continues the exploration of consciousness, shifting realities, and self-realization, promising deeper insights and reflections.

delves into multiverse theory, quantum entanglement, and parallel universes, unraveling mysteries and questioning existence's fabric.

32. **Dreamscapes: A Journey into the Parallel Universe of the Subconscious Mind**

    Dreamscapes invites readers to explore the subconscious mind. Delve into lucid dreams, emotions, and memories, revealing connections between dreaming and waking life through captivating prose and analysis.

33. **The Eternal Quest: A Journey to Enlightenment**

    The Eternal Quest explores the human pursuit of enlightenment. Through vivid storytelling, traverse consciousness landscapes and discover profound insights into the nature of existence and spiritual awakening.

34. **Enigma Unveiled**

    Enigma Unveiled takes readers on a journey through history and the supernatural. Blending folklore with mystery, explore haunted estates and cursed artifacts, revealing hidden dimensions and enigmas.

35. **Buying Time: Unleashing the Natural Timing of Money**

    Buying Time explores aligning financial decisions with natural cycles. Drawing on economics and psychology, this guide offers strategies for mindful spending, investing, and achieving financial freedom.

36. **The Complete Millionaire Activation Guide: Manifesting Infinite Wealth**

    This guide combines metaphysical principles with practical strategies to unlock wealth potential. It offers tools to align mindset, energy, and action, empowering readers to manifest abundance on every level.

37. **Ego War: The Evolution of AI, Humanity, and the Battle for Brilliance**

    Explore the profound intersection of artificial intelligence and human creativity with *Ego War: The Evolution of AI, Humanity, and the Battle for*

Brilliance. This groundbreaking book examines the synergy between AI and humanity, offering a visionary perspective on the challenges and opportunities of our shared future.

The end.

www.ingramcontent.com/pod-product-compliance
Lightning Source LLC
Chambersburg PA
CBHW072342290526
45794CB00002B/987